SpringerBriefs in Fire

W0230623

Series Editor

James A. Milke

For further volumes:
http://www.springer.com/series/10476

Newport Partners LLC

Home Fire Sprinkler Cost Assessment

 Springer

Newport Partners LLC
Davidsonville, MD, USA

ISSN 2193-6595   ISSN 2193-6609 (electronic)
ISBN 978-1-4939-1082-3   ISBN 978-1-4939-1083-0 (eBook)
DOI 10.1007/978-1-4939-1083-0
Springer New York Heidelberg Dordrecht London

Library of Congress Control Number: 2014942695

Printed on acid-free paper

Springer is part of Springer Science+Business Media (www.springer.com)

Foreword

This book updates the report, Home Fire Sprinkler Cost Assessment (Fire Protection Research Foundation 2008). The primary purpose of this study is to review current home fire sprinkler system costs against the 2008 benchmark study to gain a better understanding of how increasingly widespread adoption of sprinkler ordinances impacts system cost. Using a larger sample size, the current study attempts to gain a better understanding of the impact of sprinkler ordinances on home fire sprinkler system cost and other factors that affect system cost.

The content, opinions, and conclusions contained in this book are solely those of the author and do not necessarily represent the views of the Fire Protection Research Foundation. The Foundation makes no guaranty or warranty as to the accuracy or completeness of any information published herein.

Project Technical Panel

David Butry, National Institute of Standards and Technology
Mike Chapman, Chapman Homes
Paul Coad, Residential Sprinkler Design
Paul Emrath, National Association of Home Builders
Jeff Feid, State Farm Insurance
Tony Fleming, Metropolitan Fire Protection
Dennis Gentzel, U.S. Fire Administration
Tonya Hoover, California State Fire Marshal
Bruce Johnson, International Code Council
Gary Keith, FM Global Corporation
William Kirkpatrick, East Bay Municipal Utility District
Paul Valentine, Nexus Technical Services

Liaison

Peg Paul, Home Fire Sprinkler Coalition
Lorraine Carli, National Fire Protection Association

Principal Sponsor

National Fire Protection Association

Preface

This book updates the report, *Home Fire Sprinkler Cost Assessment* (Fire Protection Research Foundation 2008). The primary purpose of this study is to review current home fire sprinkler system costs against the 2008 benchmark study to gain a better understanding of how increasingly widespread adoption of sprinkler ordinances impacts system cost. Using a larger sample size, the current study attempts to gain a better understanding of the impact of sprinkler ordinances on home fire sprinkler system cost and other factors that affect system cost.[1]

The current study examines 51 homes in 17 communities; the 2008 study examined 30 homes in 10 communities. The long-term goal is to enlarge and repeat the study at regular intervals to measure costs within an increasing number of communities over time. Several key statistics comparing the two studies are shown in Table 1.

The 2008 report was completed prior to the International Residential Code's sprinkler requirement, *Section R313 Automatic Fire Sprinkler Systems,* applicable to townhouses and one-and-two family dwellings. The requirement has also been included in the NFPA 1 Fire Code, the NFPA 101 Life Safety Code, and NFPA 5000 Building Construction and Safety Code since the 2006 editions of each code. Section R313.2.1 of the IRC requires that fire sprinkler systems be installed either in accordance with Section P2904 (Dwelling Unit Fire Sprinkler Systems)[2] of the IRC or with the National Fire Protection Association (NFPA) 13D standard. Prior to the section's inclusion in the 2009 IRC, residential fire sprinkler provisions were previously available via appendix. However, none of the communities in the study require that systems be installed in accordance with Section P2904.[3] Instead, all jurisdictions apply the NFPA 13D standard, or NFPA 13D with local amendments.

---

[1] Variables that factor into sprinkler system costs include product and installation costs, permit and inspection fees, increased tap fees, testing and design fees, and additional equipment such as backflow preventers, booster pumps, and holding trunks where required.

[2] Section P2904 is a prescriptive residential fire sprinkler design and installation requirement consistent with NFPA 13D and is found in the 2009 and 2012 IRC plumbing requirements.

[3] 2012 International Residential Code.

Table 1 Comparison of key statistics, 2008–2013

Select key statistics	2008 Study	2013 Study
Number of communities	10	17
Number of homes	30	51
Number of homes on public water supply/stored	24/6	44/7
Number of homes with basement/slab/crawl space foundations	20/6/4	31/18/2

Table 2 Residential sprinkler system costs, 2008–2013

	2008 Cost		2013 Cost	
	$/Sprinklered ft^2 [a]	Total cost[b]	$/Sprinklered ft^2	Total cost
Mean	$1.61	$6,316	$1.35	$6,026
Median	$1.42	$5,843	$1.22	$5,000
Minimum	$0.38	$2,386	$0.81	$1,695
Maximum	$3.66	$16,061	$2.47	$21,000

[a]The term "sprinklered square feet" (sprinklered ft^2) reflects the total area of sprinklered spaces, including basements, garages, and attics where applicable. This term is used to better characterize the cost of sprinklers per unit of space which is covered by the system, especially because many of the homes have sprinklers in spaces beyond the normal living space, such as a garage
[b]"Total Cost" includes all cost including spaces beyond the normal living space, such as garages and unfinished basements, and any associated fees

The cost analysis is provided in two major sections. *Cost Analysis of Residential Sprinkler Systems* examines all 51 homes in the study and analyzes the data across homes and, where applicable, provides community comparisons. The *Individual Community Analysis* section provides individual community data and also includes qualitative data gained through interviews with community officials, builders, and fire sprinkler contractors.

In the 2013 update the average cost per sprinklered ft^2 was $1.35. Table 2 provides a comparative summary of costs for the 2008 study and the 2013 update. Study findings also include:

- CPVC piping continues to be the material of choice;
- Significantly lower costs are found in the two states having statewide ordinances, Maryland and California, than in the sample as a whole; and,
- There is no noted increase in number of homes using a multipurpose system.

Contents

Chapter 1
Introduction

The National Fire Protection Association (NFPA) first issued NFPA Standard 13D: Standard for the Installation of Sprinkler Systems in One- and Two-Family Dwellings and Mobile Homes, in 1975.[1] As of the most recent edition in 2013, there have been 12 updates to the standard reflecting the evolution of best practices, new technologies, and overall practical experience. NFPA Standard 13D and related NFPA Standard 13R[2] have evolved to optimize system costs and fire safety for residential buildings. NFPA Standard 13D applies to one- and two-family dwellings and mobile homes while 13R applies to multifamily residential buildings up to four stories.

Since the first documented community (San Clemente, CA) adopted a sprinkler ordinance in 1978, the number of communities mandating residential sprinkler systems has continued to increase.[3] Several studies documenting costs and benefits of sprinkler ordinances have been conducted, but they tend to be based locally and lack current national cost data. In 2008, the Fire Protection Research Foundation (FPRF) undertook a study to provide a national perspective on the cost of home fire sprinklers. This study, *Home Fire Sprinkler Cost Assessment* (FPRF 2008), examined installation costs for sprinkler systems in ten communities distributed throughout the United States. The study found the cost of sprinkler systems to the homebuilder, in dollars per sprinklered square foot (ft^2), ranged from \$0.38 to \$3.66, with the average cost being \$1.61 per sprinklered ft^2.

Since the study's publication, the International Code Council (ICC) voted to include a requirement for residential sprinklers in the International Residential Code (IRC). This code provision has been adopted statewide by California and Maryland, and by some local jurisdictions in other states.

---

[1] "Mobile Homes" was replaced with "Manufactured Homes" in the 1994 edition.

[2] Standard for the Installation of Sprinkler Systems in Residential Occupancies up to and Including Four Stories in Height, NFPA 13R.

[3] Date provided by Ron Coleman (Fire Chief, San Clemente, CA).

N. Partners LLC, *Home Fire Sprinkler Cost Assessment*, SpringerBriefs in Fire, DOI 10.1007/978-1-4939-1083-0_1, © Fire Protection Research Foundation 2014

Fig. 1.1 Houses built with sprinkler systems annually

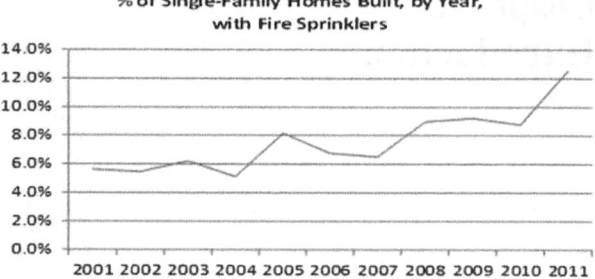

According to data from the 2009 and 2011 American Housing Survey, the total number of occupied units equipped with sprinkler systems increased over 30 % in the 2-year span following the inclusion of sprinklers in the IRC (Fig. 1.1).[4]

Although this number is increasing, only about 6 % of the current housing stock is equipped with sprinkler systems. Despite numerous reports validating the safety benefits of sprinklers, many communities refrain from mandating their use often due to cost concerns.

In some cases cost can be impacted by local ordinance and how national codes and standards are amended at the local level. For example, NFPA 13D and the IRC do not require that garages be sprinklered. However, in some jurisdictions, sprinkler ordinances are amended to require fire sprinkler coverage of garages. Amendments such as these can increase installation cost and complexity.[5]

The purpose of this study is to provide an updated review of costs for home fire sprinkler systems. Following the same methodology as the 2008 study, the current study examines a larger sample size that provides a better understanding of the relationship between sprinkler ordinance adoption and costs. The study is not a representative sample of all new homes, but the cases studied include a wide range of geography, regulations, housing types, sprinkler systems and materials, and water supply situations.

As established in the 2008 study, costs are reported primarily in sprinklered ft². Per unit cost reporting is important as it normalizes sprinkler system cost for the area served by a sprinkler system, regardless of house size or added sprinkler coverage that may be required by a local jurisdiction.

---

[4] http://factfinder2.census.gov/faces/tableservices/jsf/pages/productview.xhtml?pid=AHS_2011_S01AO&prodType=table

[5] For example, when garages are required to be sprinklered, often nonmetallic pipes in the garage require additional protection due to requirements of the IRC and in the UL product listing. Similarly, in most cases, nonmetallic pipe systems in unfinished basements also require protection. In these cases, some contractors are driven to use metallic pipe in unfinished basements and garages in order to avoid the added cost of pipe protection.

Chapter 2
Community Selection

This section contains information relating to community selection. It details the criteria for selecting communities and provides an overview of the communities identified.

To obtain builder cost data, we identified 17 communities comprising 51 homes, typically three homes per community. In contrast, the 2008 benchmark study featured ten communities (30 homes) that were selected based primarily on geography, availability of data, system type, and sprinkler ordinance status. We contacted all ten of the original communities for participation in the current study to gather data on changes in system cost over time. However, for various reasons, only five of the original communities are included in the current study. For example since 2007, Huntley, Illinois, only requires that sprinkler systems are a "mandatory option," in which builders must provide homeowners with the *option* to install a fire sprinkler system. This factor combined with reduced building activity resulted in an inability to locate sprinkler projects in that community. Four additional communities: Matteson, Illinois; San Clemente, California; Pleasant View, Tennessee; and Wilsonville, Oregon also experienced a severe slowdown of residential building activity such that we were unable to locate sprinkler system projects. In order to make up for lost data, additional similar or neighboring communities were selected. Figure 2.1 illustrates the building activity from the communities in the 2008 study as well as the new additions for 2013. Building activity is important for two reasons—first as an indicator of community size and second as an indicator of a healthy building environment relative to community size. As this study period follows a severe downturn in the housing market followed by an uneven recovery, we needed communities that had ongoing building activity in order to capture recent cost data.

Communities surveyed in 2008 that are not included in the 2013 update are highlighted in yellow.

Although some communities included in the study also experienced low building activity, these are small communities in which we were successful in locating recent buildings. These communities provided important geographic diversity to the data. The current study includes four communities from Maryland and

N. Partners LLC, *Home Fire Sprinkler Cost Assessment*, SpringerBriefs in Fire,
DOI 10.1007/978-1-4939-1083-0_2, © Fire Protection Research Foundation 2014

Community	# Single Family Building Permits since 2008
Addison, TX	53
Bakersfield, CA	4,454
Baltimore County, MD	2,748
Carroll County, MD	1,015
Cheatham and Davidson Counties, TN	6,504
Elk Grove, CA	1,374
Fort Collins, CO	1,328
Fresno, CA	4,690
Greenburgh, NY	83
Huntley, IL	573
Irvine, CA,	2,754
Kenmore, WA	307
Lake County, IL	206
Matteson, IL	27
Montgomery County, MD	4,535
North Andover, MA	81
Pitt Meadows, BC, Canada	439
Pleasant View, TN	60
Prince Georges County, MD	4,497
San Clemente, CA	139
Scottsdale, AZ	1,032
Wilsonville, OR	262

Fig. 2.1 Number of building permits since 2008

California as a result of these states having passed statewide sprinkler regulations in 2011. This data provides a better understanding of how the cost of a residential fire sprinkler system has been affected by a statewide requirement.

In addition to the effect of statewide sprinkler regulations, an effort was made to include communities that varied in terms of length of time a requirement for sprinkler systems was in effect. In total, nine communities had sprinkler ordinances for more than 5 years, six had ordinances under 5 years, and three had no ordinance.[1]

Participating communities were selected to provide geographic diversity and reflect a variety of local circumstances that impact sprinkler system cost. Differences such as type of installer (sprinkler contractor or plumber), piping materials, and local modifications to NFPA 13D vary amongst the communities. Further, having a broad geographic range allowed for the inclusion of a variety of housing types. For example, basement foundations are typical in the Northeast, while a slab foundation is more common in California and Texas.

While the status of the sprinkler ordinance and the geographic location were primary selection criteria, several other factors were evaluated with the intent of gaining a diverse set of data. For instance, at least one community with plumber-installed multipurpose systems was included, and a concerted effort was made to include a system using passive purge (Fresno, CA).

Systems cover a range of sprinkler piping materials: Chlorinated Polyvinyl Chloride (CPVC, most common), Cross-linked Polyethylene (PEX) and some black steel and copper used in combination with CPVC in basements and garages. The water supply used in the sprinkler systems was also examined to determine differences between sprinkler system costs in areas with and without a municipal water supply. The current study includes seven homes on stored water systems (well water) with the remaining homes on public water supply.

The decline of new residential building activity was a major challenge in locating communities, particularly in the Southeast. About 15 Southeast communities having sprinkler ordinances were contacted, however, lack of building activity, as well as special provisions[2] within the ordinances made it difficult to collect data relevant for this study.

The 17 communities selected for the 2013 update are illustrated on the map in Fig. 2.2.

---

[1] Cheatham County, TN and Davidson County, TN were counted as one community for the overall purpose of the study because of their proximity. However, it is important to note that while Cheatham County has a sprinkler ordinance, Davidson County does not. For this reason, we included Cheatham County in the count of homes with sprinkler ordinances over 5 years and Davidson County as having no ordinance.

[2] Many of the Southeast communities' ordinances applied only to larger homes.

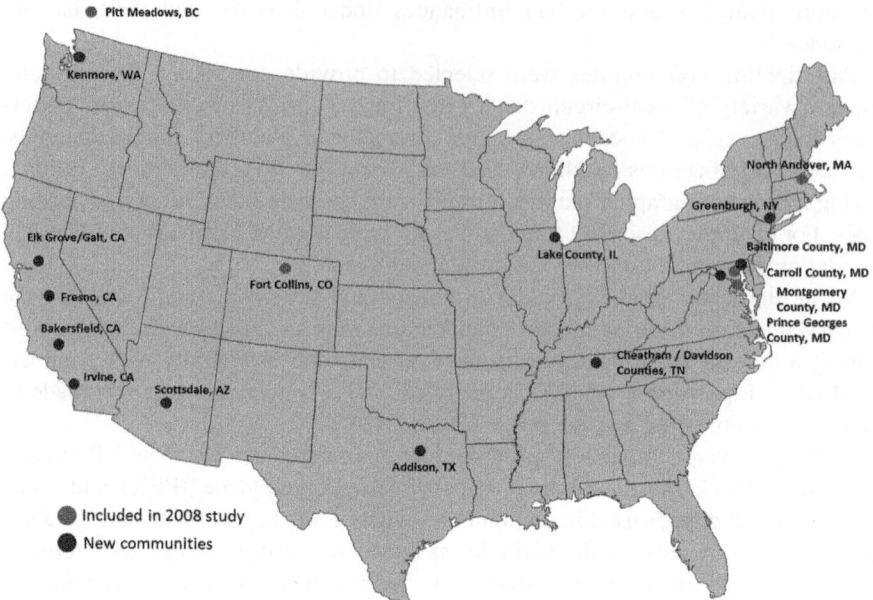

Fig. 2.2 Communities included in the 2013 study showing geographic distribution

Chapter 3
Cost Analysis of Residential Sprinkler Systems

This section discusses the separate sprinkler system costs that were analyzed and breaks down the several sprinkler system variables that factor into system cost. These variables include system requirements and extent of coverage, piping material, water source, permit and inspection fees, system design type, custom versus tract homes, foundation type, and the existence of statewide sprinkler regulations.

For each community, building cost data was requested for three homes built since 2010. In each community, house plans and cost data were typically obtained from builder or sprinkler contractors that were recommended by local fire departments or homebuilder associations. The builders and sprinkler contractors participating in the study were asked for house plans and cost data documentation, as well as supplemental information pertaining to the specific sprinkler systems and homes. Total cost data reported in this study refers to all of the cost incurred by the builder for the system and includes permit fees, increased tap fee charges, inspection fees, and any additional costs that vary by project and jurisdiction.

The house plans obtained span a cross section of housing characteristics including: tract and custom homes, area of living space, and foundation type. Additionally, sprinkler systems for each of the identified homes vary among system type; area served by the sprinkler system, the need for sprinklers in unfinished space, type of water supply, piping materials, and other system design issues such as the use of backflow preventers and recessed sprinkler heads.

In the original 2008 report, the 30 homes analyzed had an average price per ft^2 of sprinklered space of $1.61. In the 2013 study, the average cost per sprinklered ft^2 was $1.35. Increased adoption of sprinkler ordinances, improved installation methods, standardized practices, and increased contractor competition likely had an effect on reducing costs over the 5 years between studies.[1] Table 3.1 presents a comparison of the cost per sprinklered ft^2 for the 2008 study and 2013 update.

---

[1] The average total cost for the 2008 study of $6,316 declined to $6,026 in 2013. However, the total cost is dependent on house size and required sprinklered space.

N. Partners LLC, *Home Fire Sprinkler Cost Assessment*, SpringerBriefs in Fire, DOI 10.1007/978-1-4939-1083-0_3, © Fire Protection Research Foundation 2014

Table 3.1 Residential sprinkler system costs, 2008–2013

	2008 cost		2013 cost	
	$/Sprinklered ft^2	Total cost	$/Sprinklered ft^2	Total cost
Mean	$1.61	$6,316	$1.35	$6,026
Median	$1.42	$5,843	$1.22	$5,000
Minimum	$0.38	$2,386	$0.81	$1,695
Maximum	$3.66	$16,061	$2.47	$21,000

Table 3.2 Comparative costs for repeat communities[a]

	2008			2013		
Community	$/ Sprinklered ft^2	Ft^2 of sprinklered space	Total cost	$/ Sprinklered ft^2	Ft^2 of sprinklered space	Total cost
Fort Collins, CO	$3.18	4,510	$13,685	$1.83	4,950	$8,731
Pitt Meadows, BC	$1.22	2,262	$2,780	$0.94	3,239	$3,045
Prince Georges County, MD	$1.00	4,806	$4,773	$1.68	2,033	$3,263
Carroll County, MD	$2.27	3,863	$8,683	$1.23	5,130	$6,333
North Andover, MA	$1.20	4,702	$5,600	$1.34	3,828	$5,067
Total averages for all homes in repeat communities	$1.81	4,029	$7,104	$1.38	4,157	$5,900

[a]It is important to note that these figures are averages for the individual communities. Therefore, simply multiplying the average $/ft^2 by average sprinklered ft^2 will not give you the average total cost figure portrayed in the table. The total averages calculated in the bottom row of the table includes all homes in the sample, not community averages

For the five communities analyzed in both studies, results are mixed, as noted in Table 3.2. Myriad variables introduced between the homes in the 2008 and 2013 studies render conclusions about costs over time difficult to make. Although we retained five communities, in no case were we able to retain the builder and/or sprinkler contractor. Other key changes from 2008 to 2013 included: sprinkler piping material, home type (custom or production), foundation type, and water supply type.

Average overall costs declined across the homes that were analyzed in both 2008 and 2013. Three communities Pitt Meadows, BC, Fort Collins, CO, and, Carroll County, MD all show reductions in price per ft^2 of sprinklered space. Differences in sprinkler system design between the 2008 and 2013 homes for these communities, such as piping material and foundation type, impact costs in these communities. The differences in variables and cost figures for each community are discussed in further detail in the Individual Community Analysis section of this report.

Some costs increased from 2008 to 2013. In North Andover, MA and Prince George's County, MD the data shows an increase in price per sprinklered ft^2. Again, different sprinkler system variables may have contributed to a change in cost figures, which is discussed further in the later sections of the report. In summary,

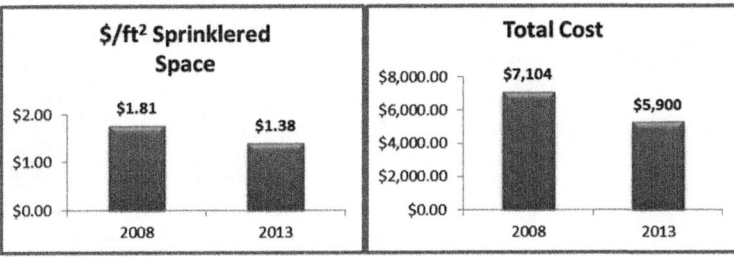

Fig. 3.1   Comparison costs for repeat communities, 2008–2013

there does not seem to be a lot to infer by comparing costs over time from the five specific communities because changes in piping material, water supply, and foundation type make the analysis inconclusive.

Figure 3.1 portrays the difference in average total cost and average price per ft^2 of sprinklered area for all communities in both studies.

3.1   Variables Impacting Sprinkler System Cost

The cost of a sprinkler system is affected by several factors. Code requirements, piping material, water supply, fees, home type (custom versus tract), and foundation type are the primary factors impacting cost. These variables are often directly related to one another. For example, sprinkler ordinances often require that unfinished areas, such as basements and garages, be covered by the sprinkler system. While most systems employ relatively inexpensive CPVC in the living area, unfinished spaces without protective covering of the sprinkler pipe are generally unsuitable for plastic piping. These mandates, therefore, essentially dictate the use of metallic piping in unfinished areas. Hence, material and installation costs are driven up for homes with basement foundations not only due to the increased area of coverage, but also due to the necessity of using more expensive piping material.

3.1.1   Sprinkler System Requirements and Extent of Coverage

In some cases local ordinances may contain sprinkler system provisions that extend beyond NFPA 13D minimum requirements. The most common extension of NFPA 13D is to require areas outside of the living space, such as garages and attics, to be covered by the sprinkler system. Subsequently, the additional space covered by sprinkler systems requires additional piping, sprinkler heads, and in some cases materials that can withstand freezing temperatures. Piping in the living space of the home is generally a plastic material (CPVC is most commonly used, but sometimes PEX), but metallic pipe is required for areas such as garages and attics. Many

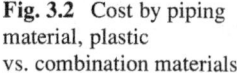

Fig. 3.2 Cost by piping
material, plastic
vs. combination materials

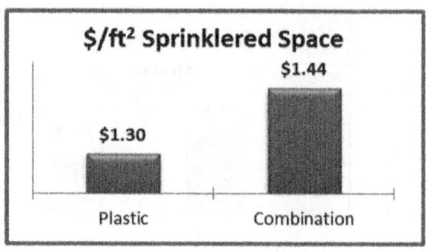

builders and contractors also choose to use metallic pipe for unfinished basements.
Adding space to be covered by the sprinkler system, particularly unfinished space in
which more expensive metallic pipe is typically installed, can significantly increase
the installed cost of the sprinkler system.

3.1.2   Type of Pipe Used

All sprinkler systems identified use either a plastic (CPVC or PEX) or a combination of
plastic and metallic (black steel or copper) piping material. CPVC is the most common
material and many systems use a combination of CPVC in the living areas and metal
piping for unfinished areas such as garages and unfinished basements. For systems
using strictly plastic piping material, average total cost of installation is \$5,866 or
\$1.30 per sprinklered ft^2. For systems that employ a mix of plastic and metal piping,
average total cost is \$6,376 or \$1.44 per ft^2 of sprinklered space (Fig. 3.2).

No homes identified for this report use strictly metallic piping materials. In 2008,
however, one community (Fort Collins, Colorado) had three systems using copper
piping which resulted in a significantly higher cost (\$3.81 per sprinklered ft^2) than
all other communities. For the current study, all Fort Collins homes had CPVC
piping at an average price of \$1.83 per sprinklered ft^2. The cost data for this
community, as well as the price figures for systems with strictly plastic piping
versus a combination of piping materials, strongly suggests that piping material is a
significant factor in sprinkler system installed cost.

3.1.3   Water Source

For the purpose of this study we designated two types of water sources: public
(municipal) and stored (homes on well water including those with a pump and tank
system). All costs associated with the sprinkler system relating to water supply (meter
upsizing, booster pump and tanks, backflow preventers, etc.) are included in the cost
figures. Most of the homes in the current study are served by a public water source.
However, seven homes from five different communities are supplied by a stored water
source. When there is a stored water source, sprinkler systems often require a booster

Fig. 3.3 Cost by water supply

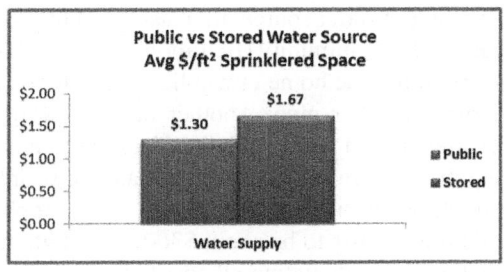

Table 3.3 Cost by water supply, 2008 vs. 2013

	2008 (24 public/6 stored)			2013 (44 public/7 stored)		
Water supply	Average sprinklered ft^2	Total cost	$/ Sprinklered ft^2	Average sprinklered ft^2	Total cost	$/ Sprinklered ft^2
Public	4,100	$5,099	$1.33	4,419	$5,918	$1.30
Stored	4,187	$11,184	$2.73	4,348	$6,706	$1.67

pump and tank, which can add significantly to the total system cost. On average, the total builder cost for systems having a stored water source is $6,706. Compared to systems that rely on a public source, with an average total cost of $5,918, a stored source increases total cost by about 13 % ($800). In terms of price per unit area of sprinklered space, stored source systems range from $1.13 to $2.44 per ft^2, with an average price of $1.67 per ft^2. The average price per ft^2 of sprinklered space for homes on public water is $1.30, 22 % lower than stored water systems (Fig. 3.3).

The difference in total cost for public and stored water supplies would have been greater had it not been for one anomalous data point. One home on public water supply (Greenburgh, NY) had a total cost of $21,000–more than $6,000 higher than the next most expensive system. By removing this home from the data, the average total cost for public water systems drops nearly $200 to $5,719 and the sprinklered ft^2 cost drops from $1.30 to $1.28.

Also notable is the drop in the cost of sprinkler systems supplied by stored water sources from the 2008 study to the present. In 2008, the average total cost of systems on stored water supply was $11,184 ($2.73 per ft^2 of sprinklered space). For the 2013 study, stored water systems average $6,706 with an average price per ft^2 of $1.67 (Table 3.3).

3.1.4   Permit, Inspection, and Additional Fees

Often a permit and/or inspection fee is required for installing a sprinkler system. These fees add to the builder's total cost. In many cases these fees were reported in the total cost figures and could not be extracted. Communities with homes supplied by a stored water source often assess additional fees for booster pumps and holding tanks. One community (Baltimore County, MD) imposed an additional flow-test fee of $650 for one home on public water supply, and required both a larger meter and larger diameter pipe. Another builder (North Andover, MA) identified separate charges for an additional electrical fee to equip systems with a bell as well as an additional plumbing cost to separate the sprinkler system from the domestic water supply at the meter. The builder who supplied the house plans estimated these additional costs to be about $300. All additional fees were factored into total costs and used in calculating all cost figures.

Some jurisdictions have no permit fees for sprinkler systems. Other communities have standardized permit fees for all homes, ranging from a low of $75 to a high of $370. Other communities use variables such as number of sprinkler heads (Montgomery County, MD) or size of home (Fort Collins, CO; Fresno, CA; Greenburgh, NY) to assess fees. All of the permit fees and other costs are included in the cost figures.

3.1.5 Backflow Preventers

Although not required by NFPA 13D, except for systems with antifreeze, many sprinkler systems are installed with testable backflow preventers. Backflow prevention devices prevent the pressurized fire sprinkler water line from flowing into the municipal water supply. Many jurisdictions require that sprinkler systems include backflow preventers in the system design. Backflow preventers for residential fire sprinkler systems typically range from approximately $75–$350.[2] Although the 2013 NFPA 13D does not require backflow preventers, it does require a test connection for backflow prevention devices. Further, the 2013 NFPA 13D allows for backflow preventers to serve as a system control valve, eliminating the need to install an additional control valve and helping mitigate the cost of adding the device (*NFPA Journal* 2012).

3.1.6 System Design Type

There are two basic types of sprinkler systems: standalone and multipurpose. A standalone system uses a dedicated sprinkler piping. A multipurpose system combines the domestic and sprinkler water supply into one common potable piping system. Standalone systems are much more common than multipurpose systems. Out of 30 homes identified in the 2008 report, only six were multipurpose systems.

---

[2] Estimate provided by a sprinkler contractor and consistent with other information found from suppliers.

Table 3.4 Foundation type market shares (http://censtats.census.gov/)

	Nationwide (US)	Northeast	Midwest	South	West
Full/partial basement	30 %	68 %	71 %	10 %	26 %
Slab	53 %	22 %	24 %	75 %	45 %
Crawl space	17 %	11 %	5 %	16 %	29 %

In the 2013 study, three multipurpose systems were identified, all in the same community (Bakersfield, CA).

The average total cost for the three multipurpose systems is $4,408 and $1.23 per ft^2 of sprinklered space. Standalone systems average $1.36 per ft^2 of sprinklered space and $6,127 total builder cost. Due to the limited data for multipurpose systems, it is not feasible to make any conclusions regarding the cost of multipurpose versus standalone systems. It is noted, however, that the 2008 study also showed lower installation costs for multipurpose systems than standalone systems.

3.1.7 Type of Foundation

Housing foundation types typically vary by geographic location. Basement foundations are typical in the Northeast and Midwest. Slab or crawl space foundations are more common in southern and western communities. Table 3.4 summarizes the market share for foundation types across the United States, according to U.S. Census data:

For this report, the average cost for sprinkler systems in homes with basement foundations is $1.44 per ft^2. Those with slab foundations average $1.14 per ft^2. Total sprinkler system cost for basement homes averages $6,730 compared to a $4,872 average for slab homes. Only two crawl space foundations were identified (Fort Collins, CO and Cheatham County, TN). These homes have an average price per ft^2 of sprinklered space of $1.71 and an average total cost of $5,503.

Figure 3.4 summarizes the price per ft^2 for sprinklered space and total cost for basement and slab foundation homes. Crawl space foundations were omitted due to the small number of crawl space homes (two) in the sample. While the average cost per ft^2 for basement foundations is higher than it is for slab foundations, there are certain variables other than foundation type that contribute to this difference. For example, many have unfinished basements which are not included in the living space but local ordinances require them to be sprinklered. As mentioned in the previous section pertaining to piping material, most communities require that sprinkler systems in unfinished areas be composed of metallic piping, which increases the cost of the system. Of the 31 basement homes, 15 had a combination of plastic and metallic piping material installed. Conversely, none of the homes with slab foundations have a combination of piping materials; all are built with plastic pipe. Builders and sprinkler contractors interviewed for this report noted the increased cost associated with metallic pipe and a strong preference for plastic.

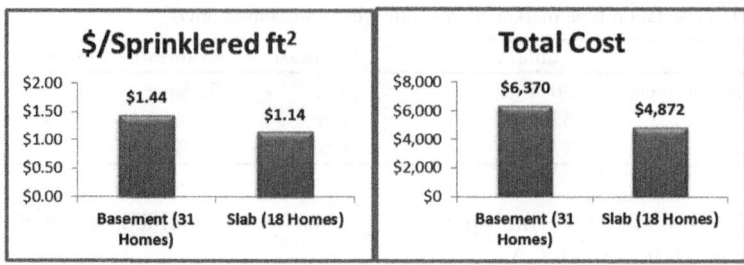

Fig. 3.4 Sprinkler system cost by foundation type

3.2   Statewide Sprinkler Requirements

Since the 2008 report, California and Maryland adopted statewide requirements for sprinkler systems in all new single-family construction. Currently, these two states are the only ones in the nation that mandate sprinkler systems statewide. Of the 17 communities in this study, eight are in Maryland and California (four from each state).

Maryland instituted a residential sprinkler system ordinance as part of the statewide adoption of the 2009 IRC effective January 1, 2011. The state requires that sprinkler systems be installed in all townhomes and one- and two-family dwellings. In the 2008 report, two Maryland communities, Prince Georges County and Carroll County, were included. In the 2013 study, the same two communities are included as are two other Maryland communities, Baltimore County and Montgomery County.

Similarly, California added a requirement for residential fire sprinklers as a result of the adoption of regulations with California specific amendments to the ICC's body of codes as part of the adoption of the 2010 California Building Standards effective January 1, 2011. The code requires residential fire sprinklers in all new one-and two-family dwellings. The 2008 report included one California community, San Clemente, which has mandated residential sprinkler systems per NFPA 13D since 1978. Data from San Clemente was unavailable for the current study, which includes four new California communities.

Comparing states with statewide sprinkler requirements to those without provides insight into the impact of statewide ordinances on system cost. Combining the 2013 data from Maryland and California, the cost for sprinkler systems is $1.16 per ft^2 sprinklered space, and $4,091 average total cost. By comparison, in all other communities the cost is $1.53 per ft^2 with a total cost of $7,877. Lower costs for systems in states with statewide requirements may be a result of more widespread

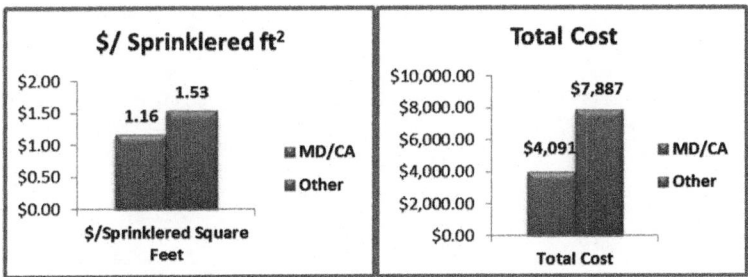

Fig. 3.5 Cost by presence of statewide ordinance (Maryland/California vs. Other)

acceptance of sprinkler systems and increased experience in installation and design practices. Market demand could also lead to lower costs as increased demand leads to competitive contractor pricing (Fig. 3.5).

Chapter 4
Individual Community Analysis

This section identifies and summarizes the different communities selected for this study. It includes background information and analyzes the data on an individual community basis. The individual community analysis is based on the data and information provided during interviews with builders, contractors, and city officials.

4.1 Addison, Texas

Addison, TX has a longstanding fire sprinkler ordinance dating back to 1992. The community, served by the Addison Fire Department, currently operates under the 2009 International Fire Code with local amendments. Some of the amendments include the requirement that sprinkler systems be installed in additions to existing homes as well as in the existing home being renovated, in the attics of all homes except those on fee simple lots, that all systems employ valves controlling water supply, and that all water flow alarm switches be electronically supervised and monitored by an approved fire alarm monitoring agency.[1] Sprinkler systems are typically installed by a sprinkler contractor and must be designed and installed in accordance with NFPA 13D standards. Standalone systems with CPVC piping are most commonly installed.

Since 2008, Addison has seen little growth in terms of new single-family construction. According to U.S. Census data, only 53 single-family building permits were issued in the past 5 years.[2]

Housing styles in Addison range between small manufactured homes and large custom homes. The average size of homes in the community is approximately 3,000 ft^2. Homes are typically built on slab foundations.

---

[1] http://www.addisontx.gov/repository/unmanaged_content/forms/Fire%20Department/Fire_Prevention_files/Amendments%20to%20the%202009%20International%20Fire%20Code.pdf

[2] http://censtats.census.gov/

Table 4.1 Community
sprinkler system costs—
Addison, TX

| | Sprinklered space | |
Total cost	Size (ft^2)	$/ft^2
$11,972	7,045	$1.70

Due to the lack of new construction in Addison, it was difficult to identify house plans for this study. As a result, information for only one home was obtained from a local sprinkler contractor. Although there is a lack of data for this community, it was important to include it in the study because of its contribution to the geographic diversity of the data. The home we identified was a 7,045 ft^2 custom home on a slab foundation with a standalone CPVC sprinkler system supplied by public water. The total cost for this system was $11,972 ($1.70 per ft^2 sprinklered space) (Table 4.1).

4.2   Bakersfield, California

Bakersfield, CA requires residential fire sprinklers as a result of the statewide adoption of the 2009 IRC as amended to be the 2010 California Residential Code (CRC). The CRC also requires fire sprinklers in attached garages. Bakersfield, served by the Bakersfield Fire Department, has experienced significant growth in terms of single family building permits, with 4,545 permits issued since 2008 according to U.S. Census Bureau figures.[3] The state requires that sprinkler systems be installed by fire protection contractors, general manufactured housing contractors, or owner-occupied owner-builders. In Bakersfield, sprinkler systems are typically installed by a sprinkler contractor. Homes are typically built on "on-grade slab" foundations and range from 2,500 to 4,000 ft^2. Sprinkler systems generally are connected to the public water source. PEX is the most common piping material for the systems.

For this study, three house plans were obtained from a sprinkler contractor. Each home was built on an on-grade slab foundation with living space ranging from 2,525 to 3,187 ft^2. In terms of sprinkler area, because of the requirement for garages, the homes range from 3,216 to 4,013 ft^2. Total builder cost for the homes ranges from $3,930 to $5,180, which includes a $180 permit fee. Cost per ft^2 of sprinklered space ranges between $1.16 and $1.29.

All three homes are equipped with a multipurpose PEX sprinkler system with concealed sprinkler heads in the living space and semi-recessed heads in the garage. Total costs are presented in Table 4.2.

---

[3] http://censtats.census.gov/

Table 4.2 Community sprinkler system costs— Bakersfield, CA

	Total cost	Sprinklered space	
		Size (ft^2)	$/ft^2
House 1	$5,180	4,013	$1.29
House 2	$3,930	3,216	$1.22
House 3	$4,115	3,537	$1.16

4.3   Baltimore County, Maryland

Baltimore County, MD instituted an NFPA 13D ordinance as part of the statewide adoption of the 2009 IRC effective January 1st, 2011. Since 2008 there has been a significant amount of new construction in Baltimore County. According to U.S. Census figures, 2,748 single-family building permits have been issued in the past 5 years.[4] Sprinkler contractors typically install residential sprinkler systems. Generally, standalone systems are installed using CPVC piping, although some systems use a combination of CPVC and black steel. Housing types in Baltimore County vary from manufactured homes to large custom homes close to 5,000 ft^2. Homes are typically built on basement foundations which are included in the sprinklered space of the home.

The home plans examined in this study were obtained from a sprinkler contractor in Baltimore County. All three homes are built on finished basement foundations so the living and sprinklered space are equal. The three homes range from 3,100 to 4,800 ft^2 of sprinklered space. Total builder cost for the three homes ranges from $3,550 to $5,225 ($1.05–$1.65 per ft^2). All three systems use semi-recessed sprinkler heads throughout (Table 4.3).

A few differences between the systems in the three homes are notable when examining total costs. In two homes (House 1 and House 2), sprinkler systems are supplied by a public water source, while House 3 employs a stored water source (well water). The well water sprinkler system in House 3 incurred the highest builder cost despite being only 50 ft^2 larger than the home having the lowest builder cost (House 2). Part of the higher total cost is attributed to a $650 flow-test fee required for one of the public supply homes.

In addition to the difference in water systems, piping material may have factored into the per unit area cost for the systems served by public water. The system at House 2 includes a combination of CPVC and black steel piping material, which may factor into the higher price per sprinklered ft^2 for House 2 over that for House 1. The two systems supplied by public water also have backflow preventers while the well water home does not (Fig. 4.1).

---

[4] http://censtats.census.gov/

Table 4.3 Community sprinkler system costs— Baltimore County, MD

	Total cost	Sprinklered space Size (ft^2)	$/ft^2
House 1	$5,050	4,800	$1.05
House 2	$3,550	3,100	$1.15
House 3	$5,225	3,150	$1.65

Fig. 4.1 Cost by water source—Baltimore County, MD

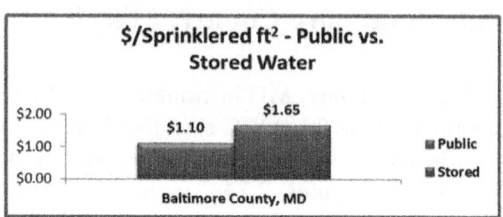

4.4 Carroll County, Maryland

Carroll County, MD is one of the communities for which data was available for both the 2008 report and the current study. Carroll County has had a sprinkler ordinance since 2006. The county, served by local paid and volunteer fire departments, has issued 1,015 single-family building permits since 2008.[5] Standalone sprinkler systems are the most common system installed; however, multipurpose systems are also allowed. CPVC piping is commonly found in the finished areas of homes and copper is typical in unfinished areas. Average home size in Carroll County is approximately 1,800 ft^2 for one-story homes, and 3,500 ft^2 for two-story homes. Basement foundations are typical throughout the county.

For the 2013 study, a local builder supplied three house plans. Living area of these plans ranges from 2,955 to 5,324 ft^2. Basement areas, including unfinished space, in Carroll County must be served by a sprinkler system. The three homes all have basement foundations with sprinklered space ranging from 4,412 to 5,729 ft^2. All three systems have CPVC piping in finished areas and copper in the unfinished basement areas. The systems are all supplied by a stored water source, have semi-recessed sprinkler heads, and include a backflow preventer.

Referencing the 2008 study, Carroll County was one of the more expensive communities in terms of costs of sprinkler systems in both total cost and price per ft^2. The homes had an average total cost of $8,683 ($2.27 per ft^2). In the 2013 update, total cost of sprinkler systems ranged from $5,000 to $7,000, averaging $6,333. Price per sprinklered ft^2 ranged from $1.13 to $1.33, averaging to $1.23 (Table 4.4).

Permit fees are included in the total cost figures. Additionally, because all the homes rely on well water, a booster pump and tank required for the systems are included in the total costs.

---

[5] http://censtats.census.gov/

Table 4.4 Community sprinkler system costs— Carroll County, MD

| | Total cost | Sprinklered space | |
		Size (ft^2)	$/ft^2
House 1	$5,000	4,412	$1.13
House 2	$7,000	5,729	$1.22
House 3	$7,000	5,248	$1.33

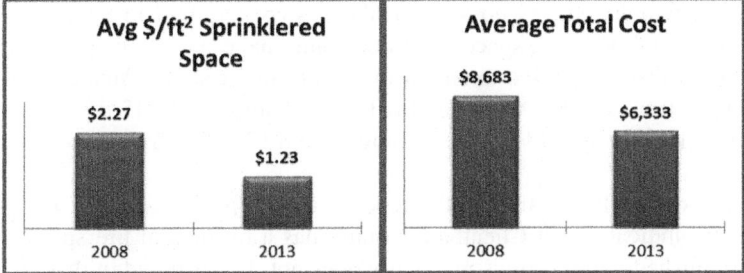

Fig. 4.2 Carroll County cost comparison, 2008–2013

Figure 4.2 depicts the difference in price, both in total cost and price per ft^2, of sprinklered space for Carroll County between 2008 and 2013. Because system type, piping material, and water supply for all homes are the same, the price decrease was attributed to efficiency in design, standardization in installation and manufacturing practices, and greater acceptance amongst professionals.

4.5   Cheatham and Davidson Counties, Tennessee

Cheatham County was the first county in Tennessee to require residential sprinkler systems for single-family homes in subdivisions of more than three lots, effective August 2006.[6] Two cities in Cheatham County (Pleasant View and Ashland City) passed sprinkler requirements in 2001 and another (Kingston Springs) passed an ordinance in 2005. Pleasant View was one of the communities analyzed in the 2008 report. Davidson County does not currently have a sprinkler system requirement although the installation of sprinkler systems remains prevalent throughout the community.[7]

In both counties, sprinkler systems are being installed and designed in accordance with NFPA 13D standards. Standalone systems are the most common system found in these communities, generally equipped with CPVC piping. Housing styles

---

[6] Tennessee Department of Commerce and Insurance, May 2010, "Cost Effectiveness of Fire Sprinkler Equipment."

[7] http://www.tn.gov/fire/documents/firesprinklerstudytmha51410.pdf

range from one and two-story homes between 1,200 and 4,000 ft², to large custom homes over 6,000 ft². Foundation type varies among homes in the area.

For this study, three homes were obtained from a local sprinkler contractor. Two were located in Cheatham County and the other in Davidson County. The size of the homes ranges from 2,400 to 8,245 ft². Two of the homes have basement foundations and attached garages that are required to be served by the sprinkler system, making the area of sprinklered space greater than the living space. The third home has a crawl space foundation and no additional areas are required to be served by the system outside of the living space. All three homes have standalone CPVC systems. No permit fees were assessed to any of the homes in the study. Area of sprinklered space ranges from 2,400 ft² for the crawl space home to 11,045 ft². Total builder cost of installing the sprinkler systems ranges from $2,640 to $12,700 ($1.10–$1.35 per ft²).

As was previously mentioned, the report, *"Cost Effectiveness of Fire Sprinkler Equipment"* indicates that Cheatham County has a residential fire sprinkler ordinance established while Davidson County does not. It was noted in that report that one of the major obstacles for sprinkler systems in Davidson County is getting water supply to the home. The report states that, "Contractors in Davidson County have said that in an average-sized home more than half the cost of the sprinkler system is the upgrades in tap fees, larger meters and backflow prevention devices."[8] As portrayed in Table 4.5, the home in Davidson County (House 2) had the highest price per sprinklered area. This data is consistent with the information retrieved from the aforementioned report. As a result it can be assumed that the increased price per unit area costs can be related to costs associated with the home's water supply (Fig. 4.3).

4.6   Elk Grove, California

The city of Elk Grove is served by the Cosumnes Community Services District Fire Department, a merger of the Elk Grove Community Services District and the Galt Fire Protection District. Since 2008, 1,374 single-family building permits were issued in the community, a significant number for a city with a population of 154,908.[9] State law limits the installation of sprinkler systems to fire protection contractors, general manufactured housing contractors, and owner-occupied owner-builders. In Elk Grove, sprinkler systems are most commonly installed by a sprinkler contractor. Homes are generally built on "on-grade slab" foundations and typically range between 2,000 and 5,000 ft². Sprinkler systems are usually connected to a public water supply and the most common piping material is CPVC.

[8] http://www.tn.gov/fire/documents/firesprinklerstudytmha51410.pdf

[9] http://censtats.census.gov/

Table 4.5 Community
sprinkler system costs—
Cheatham and Davidson
Counties, TN

	County	Total cost	Sprinklered space Size (ft^2)	$/ft^2
House 1	Cheatham	$2,640	2,400	$1.10
House 2	Davidson	$6,480	4,800	$1.35
House 3	Cheatham	$12,700	11,045	$1.15

Fig. 4.3 Cheatham and
Davidson Counties, TN—
cost breakdown

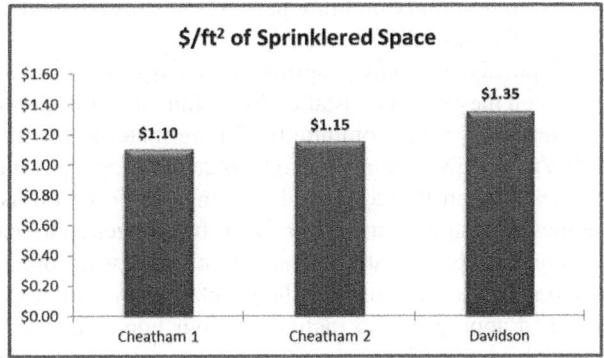

For the purpose of this study, data for three homes was collected from a sprinkler contractor. Each home is built on an "on-grade slab" foundation and ranges from 2,004 to 4,258 ft^2. Each home has a standalone system connected to the public water supply. Sprinklers are equipped with a passive purge and looped connection system with a remote toilet connected to the system for water circulation. Systems in each of the selected homes are constructed of CPVC piping.

The price per ft^2 for the three homes ranges from $0.85 to $1.11, with the price decreasing as the house size increases. Total costs for the homes range from $2,229 to $3,600.

A fee of $370 is included in the total costs to builder for plan review and inspection. It is important to note that the $370 fee is a base fee for the first inspection of each floor plan type. Each additional home of the same floor plan is charged an hourly rate for inspection. Elk Grove has worked with local water departments to keep connection fees from being increased due to the need for a larger connection for the sprinkler system (Table 4.6).

4.7   Fort Collins, Colorado

Fort Collins, CO was analyzed in both the 2008 study and the 2013 update. Communities in both studies allow for comparisons in price, material, and other innovations over the 5-year period. Fort Collins, which has mandated NFPA 13D since 1986, is served by the Poudre Fire Authority. Since the original study in 2008, Fort Collins has experienced a steady increase in new residential construction.

Table 4.6 Community sprinkler system costs—Elk Grove, CA		Sprinklered space	
	Total cost	Size (ft^2)	$/ft^2
House 1	$2,229	2,004	$1.11
House 2	$3,332	3,358	$0.99
House 3	$3,600	4,258	0.85

1,328 single-family building permits have been issued in the past 5 years according to U.S. Census data.[10]

Typically, residential sprinkler systems are installed by sprinkler contractors, although they may be installed by a plumber. Multipurpose and standalone systems are installed in the community. Piping material for the systems is usually plastic (CPVC or PEX), but may also be metallic (copper) which was the case in the three homes used in the 2008 study. Homes in Fort Collins range from manufactured homes to custom homes over 5,000 ft^2 and generally have basement foundations.

For the updated study four house plans were obtained from a local sprinkler contractor. Because the four house plans vary with respect to system design, it was deemed appropriate to include all four homes in the study. All four systems are standalone, use CPVC piping, and have concealed sprinkler heads. Homes range in size from 1,800 to 6,600 ft^2. Three of the four homes have basement foundations while one home (House 2) is built on a crawlspace foundation. The three homes with basement foundations were built to comply with 13D and there was no local requirement for sprinklers in the basement. The home with the crawlspace foundation has a sprinklered area greater than the living area because the crawlspace is also used for storage and, therefore, was required to be covered by the sprinkler system. Price per ft^2 of sprinklered space ranges from $1.55 to $2.32. Total costs to builder for the homes ranged from $8,365 to $10,200 (Table 4.7).

For each of the homes, a permit fee is included in the total system cost. Permit fees vary based on home size. One interesting note is that two of the homes are supplied by a public water source and include backflow preventers. The other two homes are supplied by a stored water source and do not require a backflow preventer because the water is a "separate supply." The well homes use an autofill system with an air gap into the water tank. The chart below compares the average price per ft^2 of sprinklered space of the four systems. It is important to keep in mind that one of the homes on well water (House 2) is also on a crawlspace foundation. Regardless, there is a clear difference in pricing per unit area between the two types of water supply (Fig. 4.4).

In the 2008 study, Fort Collins had the highest system costs among all communities analyzed, with an average total sprinkler system cost of $13,685–almost $4,000 more than any other community. In terms of sprinklered area, systems in Fort Collins cost over $1.00 more per ft^2 than the next highest community. A major factor in the high pricing was the copper piping material used in the sprinkler

---

[10] http://censtats.census.gov/

Table 4.7 Community sprinkler system costs—Fort Collins, CO

	Total cost	Sprinklered space	
		Size (ft^2)	$/ft^2
House 1	$7,385	4,100	$1.80
House 2	$8,365	3,600	$2.32
House 3	$8,975	5,500	$1.63
House 4	$10,200	6,600	$1.55

Fig. 4.4 Fort Collins, CO cost by water supply

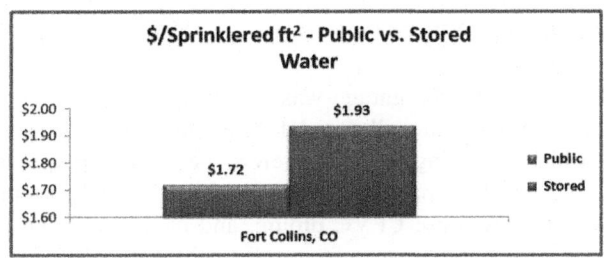

systems. Homes in the current study have sprinkler systems equipped with CPVC piping instead of copper. Average total cost of the systems in the 2013 study is $8,731, almost $5,000 lower than the 2008 average total cost. Average price per ft^2 of sprinklered space was also dramatically reduced from $3.81 to $1.83, a decrease of nearly $2.00. According to the sprinkler contractor who supplied the 2013 data, CPVC piping is the most common material used in Fort Collins at the present time. Further, all the systems studied in 2008 were on a stored water supply, which can significantly increase costs. The 2013 study includes two homes on stored water and two homes on public water (Fig. 4.5).

4.8   Fresno, California

The city of Fresno, CA adopted a residential fire sprinkler requirement as a result of the statewide adoption of the 2009 IRC and 2010 CRC. The CRC also requires fire sprinklers in attached garages. There are no additional local modifications to NFPA 13D. The city is served by the Fresno Fire Department. Sprinkler systems are typically installed by a sprinkler contractor. Since 2008, Fresno has experienced substantial growth in new residential construction. Over the past 5 years, 4,690 single-family building permits have been issued in the city according to U.S. Census data.[11] The typical house in Fresno, CA is a ranch style single story home built on a slab foundation ranging between 2,000 and 3,000 ft^2.

Three house plans were obtained from a builder for the purpose of this study. Each home is built on a slab foundation and ranges in size from 1,694 to 2,223 ft^2.

---

[11] http://censtats.census.gov/

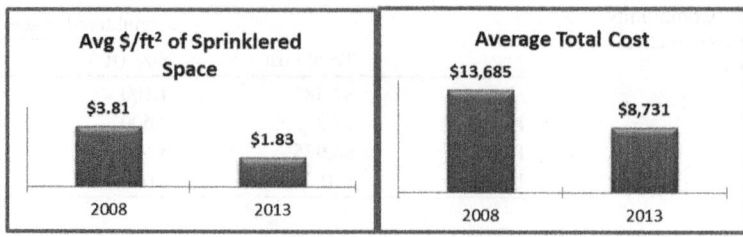

Fig. 4.5 Fort Collins, CO cost comparison, 2008–2013

In each home the garage was required to be served by the sprinkler system. The three systems are all standalone systems using a passive purge system.[12] The builder supplying the data prefers the passive purge and only uses the backflow preventer if the passive purge is not allowed. All systems are connected to a public water source, use CPVC piping, and have concealed sprinkler heads. The sprinklered area of the homes ranges from 2,090 to 2,647 ft^2. Total cost to builder for the systems ranges from $1,695 to $2,252. This includes additional permit and inspection fees, which vary among the three homes but not by a significant amount. The average price per ft^2 of sprinklered space is $0.85 (Table 4.8).

It is important to note that the four California communities selected for this report all have cost figures that are well below the average costs in the study. Three of the communities (Elk Grove, Irvine, and Fresno) all have average price per ft^2 of sprinklered space figures under $1.00. In all of these communities, standalone systems are installed in the homes. The other California community (Bakersfield) has a price per ft^2 average of $1.23 for multipurpose systems. Only one other community (Pitt Meadows, BC) has an average price per ft^2 of sprinklered space under $1.00.

4.9   Greenburgh, New York

Greenburgh, NY adopted a residential fire sprinkler ordinance in 1994. The ordinance, amended in 2011, requires that sprinkler systems comply with NFPA 13D standards. The town is served by three independent fire districts (Fairview, Greenville, and Hartsdale). Residential sprinkler systems are installed either by a sprinkler contractor or plumber. Sprinkler systems are generally standalone systems that use a combination of plastic (CPVC) and metallic (black steel) piping materials. Since 2008, only 83 single-family building permits have been issued in the community according to U.S. Census figures.[13] Homes in Greenburgh, NY are typically custom homes ranging from 1,500 to over 6,000 ft^2.

---

[12] NFPA 13D defined a passive purge sprinkler system as, "a type of sprinkler system that serves a single toilet in addition to the fire sprinklers."

[13] http://censtats.census.gov/

Table 4.8 Community
sprinkler system costs—
Fresno, CA

		Sprinklered space	
	Total cost	Size (ft^2)	$/ft^2
House 1	$1,695	2,090	$0.81
House 2	$2,252	2,399	$0.94
House 3	$2,181	2,647	$0.82

Table 4.9 Community
sprinkler system costs—
Greenburgh, NY

		Sprinklered space	
	Total cost	Size (ft^2)	$/ft^2
House 1	$21,000	8,500	$2.47
House 2	$11,000	6,000	$1.83
House 3	$8,000	4,500	$1.78

For this study three house plans were obtained from a sprinkler contractor. All three homes are built on basement foundations, thus requiring sprinkler coverage in the basement in addition to the main living areas. Like the majority of homes built on basement foundations, the systems used a combination of CPVC and black steel[14] for piping materials. CPVC is installed in the living areas and black steel is installed in basements and attached garages. Living space of the three homes ranges from 3,000 to 6,000 ft^2. Sprinklered space ranges from 4,500 to 8,500 ft^2 with an average price per ft^2 of $2.03. The average total cost of the sprinkler systems for the three homes is $13,333 and ranges from $8,000 to $21,000 (Table 4.9).

For this community we were able to separate material and installation costs for the sprinkler systems. Material cost averages $4,352 for the three systems while the average installation cost is $8,650. Additionally, design and permit fees range from $220 to $450.

4.10   Irvine, California

Irvine, CA instituted a residential fire sprinkler requirement in 2011 as a result of the statewide adoption of the 2009 IRC and 2010 CRC. The CRC also requires fires sprinklers in attached garages. The statewide requirement mandates sprinkler systems to be designed and installed in accordance with NFPA 13D standards, or the prescriptive option found in the CRC. As required by state law, sprinkler systems must be installed by a fire protection contractor, general manufactured housing contractor, or owner-occupied owner-builder. Irvine requires a single, inline check valve at the point-of-connection to the domestic water supply at the

---

[14] Black steel, black iron and steel are terms used by contractors in collecting data for this report. It is all the same product and referred to within the report as black steel, unless it is galvanized steel which is noted separately.

dwelling for backflow prevention. Irvine is served by the Orange County Fire Authority. Since 2008, the community has experienced a substantial amount of new single-family construction with 2,754 single-family building permits issued in the past 5 years according to the U.S. Census Bureau.[15] Standalone systems using CPVC piping material are most common; housing styles vary between tract and custom homes; sizes range from 2,000 to over 6,000 ft^2, and are typically built on slab foundations.

For the current study, a local sprinkler contractor supplied four house plans. The homes employ CPVC piping equipped with concealed sprinkler heads. All of the homes are supplied by a public water source. The average size of the homes is 3,688 ft^2, ranging from 2,570 to 6,240 ft^2. Like most homes on slab foundations, sprinklered space is equal to living space for each home. Installed system costs ranged from \$0.87 to \$1.44 per ft^2 of sprinklered space, averaging \$1.03 per ft^2. Total builder costs range from \$2,408 to \$9,008, averaging \$4,167 (Table 4.10).

It is important to note that the three tract homes (House 1, House 2, House 3) have similar cost figures. The custom home (House 4) is the largest and has the highest cost (on the basis of total and per ft^2 cost). Removing House 4 from the data results in an average price per ft^2 of sprinklered space of \$0.90 (\$2,522 total cost). Additionally, for this community we were able to extract material costs and additional fees (permit and inspection) from the total cost. The three tract homes incurred a standard permit fee of \$215 and have an average materials cost of \$1,285. The permit fee for the custom home was \$650 and materials cost is \$5,000. According to the sprinkler contractor providing the data, increased cost for the custom home was driven by higher overhead and labor costs for a home not on a standard plan. The chart below portrays the difference in price per ft^2 of sprinklered space between the three tract homes and the custom home (Fig. 4.6).

4.11 Kenmore, Washington

Kenmore, WA adopted a residential fire sprinkler ordinance in 2012 requiring sprinkler systems be designed and installed in accordance with NFPA 13D standards. Kenmore is served by the Northshore Fire Department. According to U.S. Census figures, 307 single-family building permits have been issued in the community since 2008.[16] Although low compared to other communities in this study, this number shows steady growth for a population of only 20,000. Sprinkler systems in Kenmore are typically installed by a sprinkler contractor. Typical housing styles in Kenmore are one or two story homes ranging from 1,500 to over 3,000 ft^2. Kenmore requires that only the living space be covered by the

---

[15] http://censtats.census.gov/

[16] http://censtats.census.gov/

Table 4.10 Community sprinkler system costs—Irvine, CA

	Total cost	Sprinklered space Size (ft^2)	$/ft^2
House 1	$2,408	2,570	$0.94
House 2	$2,466	2,830	$0.87
House 3	$2,784	3,112	$0.89
House 4	$9,008	6,240	$1.44

Fig. 4.6 Irvine, CA cost by home type

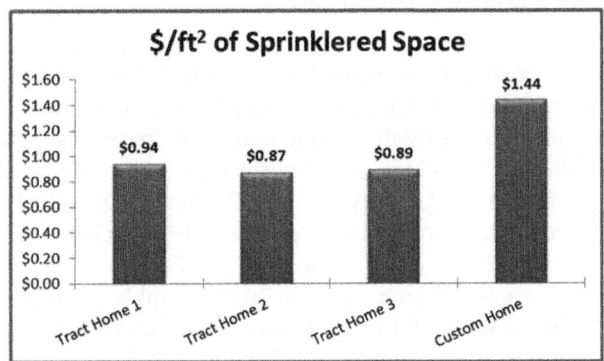

sprinkler system, which in the case of the homes in this community includes finished basements.

For this report three house plans were obtained from a local sprinkler contractor. The three homes have "two-story" foundations. The contractor who provided the data explained the homes are built on a slope, with the garage and half of the basement being slab on grade and the other half being a walkout basement. For the purpose of this study we categorized the homes in this community as having basement foundations because the characteristics most closely resemble other basement homes. All three homes are supplied by a public water source, use CPVC piping for the sprinkler systems, and are equipped with concealed sprinkler heads.

The size of the three homes in Kenmore range from 2,498 to 2,601 ft^2. Total costs to builder for the installation of sprinkler systems range from $4,540 to $4,740. This cost includes a flat permit fee of $240 for each home. Interestingly, the normalized cost among the three homes is identical at $1.82 per ft^2 of sprinklered area. Two homes have the same floor plan resulting in identical data. Additionally, because the sprinkler ordinance for this community only requires that living space be covered by the system, square footage for sprinklered space and living space is identical (Table 4.11).

Table 4.11 Community
sprinkler system costs—
Kenmore, WA

	Total cost	Sprinklered space	
		Size (ft²)	$/ft²
House 1	$4,740	2,601	$1.82
House 2	$4,540	2,498	$1.82
House 3	$4,540	2,498	$1.82

4.12   Lake County, Illinois

Lake County, IL does not have a county-wide residential fire sprinkler ordinance, but jurisdictions within the county have adopted sprinkler ordinances. Towns and villages served by the Countryside Fire Protection District are required to install residential sprinkler systems since an ordinance was adopted in 2004. Some portions of the Countryside Fire Protection Districts have had fire sprinkler ordinances as far back as 1988 (portions of Long Grove are served by the Countryside District). Some jurisdictions in Lake County have only recently added an ordinance, such as Vernon Hills in 2012. Sprinkler systems are required to be designed and installed per NFPA 13D standards. The homes in this study are not located in the Countryside Fire Protection District, but are located within Lake County. However, homes have been built in other parts of Lake County with residential fire sprinklers, possibly due to the proximity to districts with fire sprinkler ordinances.

Since 2008, 206 single-family building permits have been issued in Lake County, IL, according to U.S. Census figures.[17] The systems are generally supplied by a public water source and use CPVC piping material. Typical housing styles in Lake County are two-story single family homes ranging from 2,000 to over 6,000 ft².

For this study three house plans were obtained from a sprinkler contractor in Lake County. The three homes are built on basement foundations and basements are required to be covered by the sprinkler system. The basements are all finished space allowing for CPVC to be used. Each home also includes a garage with a "warm wall, sidewall sprinkler" meaning that the wall was insulated with no pipes actually entering the garage. The sprinkler systems are standalone systems supplied by a public water source and employ CPVC piping material throughout.

Size of the homes ranges from 3,828 to 7,890 ft² with sprinklered area ranging from 4,084 to 8,146 ft². Total builder installation cost ranges from $8,491 to $13,909 and includes a standard $200 permit fee for all homes. Price per sprinklered ft² in Lake County ranges from $1.71 to $2.08. It is interesting to note that the price per unit area declines as the total area increases. These are all custom homes so the price decline is likely a result of the design costs being spread over a larger area (Table 4.12).

---

[17] http://censtats.census.gov/

Table 4.12 Community sprinkler system costs—Lake County, IL		Total cost	Sprinklered space	
			Size (ft^2)	$/ft^2
	House 1	$9,386	5,121	$1.83
	House 2	$8,491	4,084	$2.08
	House 3	$13,909	8,146	$1.71

4.13 Montgomery County, Maryland

Montgomery County has a history with residential sprinkler systems dating back to the 1990s for townhomes. The sprinkler coverage was extended over the years with incentives for existing homes through a property tax credit worth up to half of the property tax in the year of installation effective July 1, 2000. In October 2003, Montgomery County, MD passed a bill requiring that sprinkler systems be installed in all new single family construction. The law, which went into effect January 1, 2004, made Montgomery County the largest jurisdiction in the nation to mandate sprinkler systems in new homes.[18]

The current statute requires that sprinkler systems be installed per NFPA 13D standards. It also requires that garages be covered by the system if they are wholly or partially under the living space. Sprinklers are also required under attached balconies or porches unless more than half of the longest exterior side is open to the atmosphere. Standalone sprinkler systems are commonly installed in the community using a combination of plastic and metallic piping materials. The county is served by the Montgomery County Fire and Rescue Service, which includes a combination of 40 career/volunteer fire stations.

Montgomery County is the most populous jurisdiction in Maryland with over 1 million people. Since 2008, the county has experienced a substantial increase in new construction with 4,535 single-family building permits issued. A wide variety of housing styles are available ranging from condominiums and townhomes to modest single-family and luxury homes. As a result, home sizes range from 1,000 to over 6,000 ft^2.

For this report, three house plans were obtained from a local builder. Homes range from 4,074 to 5,268 ft^2, including the basement which is required to be covered by the sprinkler system. The three homes are all built with basement foundations and served by a public water source. Sprinkler systems are all standalone using a combination of CPVC and copper piping. Total cost to builder for installing the systems ranges from $4,517 to $5,130. This includes permit and inspections fees. Fees vary based on number of sprinkler heads in the system design, at $5.30 per head. Price per ft^2 of sprinklered space ranges from $0.92 to $1.26.

Compared to the other three Maryland communities identified in this study, Montgomery County has the lowest price per ft^2 of sprinklered space. This can be

---

[18] http://www6.montgomerycountymd.gov/Apps/firerescue/press/PR_details.asp?PrID=2624

Table 4.13 Community sprinkler system costs— Montgomery County, MD		Sprinklered space	
	Total cost	Size (ft^2)	$/ft^2
House 1	$4,834	5,268	$0.92
House 2	$4,517	4,896	$0.92
House 3	$5,130	4,074	$1.26

attributed at least in part to a longstanding sprinkler ordinance along with high building volume in the county. Better installation practices and overall knowledge and comfort with the systems as well as market demand likely contributed to the lower costs in Montgomery County compared with the rest of the state (Table 4.13).

4.14   North Andover, Massachusetts

North Andover, MA is one of the communities analyzed in both the 2008 and 2013 reports. As was the case in 2008, the community does not require residential sprinklers by law. However, if a sprinkler system is installed it is required to adhere to NFPA 13D standards. In addition to NFPA 13D, the North Andover Fire Department requires sprinkler systems to cover the garage area of the home. Standalone systems are generally installed in the community using CPVC piping or a combination of CPVC and metallic piping in basement and garage areas. Typical housing in North Andover is one- and two-story homes on basement foundations ranging from 2,000 to 3,500 ft^2. Since 2008, 81 new single-family building permits have been issued in North Andover according to the U.S. Census Bureau.[19]

For this study, three house plans were obtained from a local builder. The homes are built on basement foundations with living space ranging from 1,800 to 2,650 ft^2. When adding the basement and garage areas, which are required to be covered by the sprinkler system, the homes range from 2,864 to 4,314 in sprinklered ft^2. The total builder cost for the homes ranges from $4,200 to $6,800 with cost per sprinklered ft^2 ranging from $1.11 to $1.47 (Table 4.14).

All three sprinkler systems are standalone systems on public water supply using a combination of CPVC and black steel piping material. CPVC piping is installed in the living areas and black steel in the basement and garage. The builder providing the cost figures noted that the black steel was a major cost factor estimating the impact to about one-third of the total system cost. In addition, the sprinkler heads in the garage are required to be frost protected and include a plunger system. Total costs include a $150 permit fee and an estimated $300 for additional electrical and plumbing work. The builder noted that the electrical work included the need to install a bell that is connected to the flow switch of the system. The additional

---

[19] http://censtats.census.gov/

Table 4.14 Community sprinkler system costs—North Andover, MA

	Total cost	Sprinklered space	
		Size (ft^2)	$/ft^2
House 1	$4,800	4,314	$1.11
House 2	$4,200	2,864	$1.47
House 3	$6,200	4,306	$1.44

plumbing cost is needed to "T-off" the system at the water meter so that only the domestic water is metered.

It is interesting to note that when comparing the data from the 2008 report to that of the current study, North Andover costs actually increase in regards to price per sprinklered area. In 2008 the average price was $1.20 per ft^2 sprinklered area compared to $1.34 in 2013. However, the increase in cost can most likely be attributed to the use of black steel piping material in the basement and garage areas of the homes in 2013. The homes in the 2008 study used solely CPVC piping.

4.15 Pitt Meadows, British Columbia, Canada

Pitt Meadows has mandated residential fire sprinklers per NFPA 13D standards since 1998. Served by the Pitt Meadows Fire Department, it is the only community in this study located outside of the U.S. It was analyzed in both 2008 and 2013. Standalone and multipurpose sprinkler systems have been installed in the community. The typical housing style in Pitt Meadows is a two-story, 2,500 ft^2 home on a slab or crawlspace foundation. Since 2008, 435 single-family building permits have been issued.

For the 2013 study, three house plans were obtained from a local builder. The three homes are tract homes that all use the same floor plan. The homes are built on a basement foundation, which is required to be covered by the sprinkler system. Each home has a standalone sprinkler system supplied by a public water source. The piping material for the systems is CPVC. The systems also include a backflow preventer. Each home has 3,239 ft^2 sprinklered area including living space and unfinished basement. Total builder cost is $3,045 ($0.94 per ft^2 sprinklered area). This cost includes a 5 % government tax on sprinkler systems (Table 4.15).

In 2008, the average total sprinkler system cost in Pitt Meadows, BC, was $2,780 ($1.22 per ft^2). The three homes analyzed in the 2008 study were built on slab foundations, resulting in the same area of sprinklered and living space. The homes examined in the current study are built on basement foundations, thus requiring more area covered by the sprinkler system. As a result, total cost of the sprinkler systems is higher in 2013 than it was in 2008, however the price per ft^2 of sprinklered space decreased by 23 %. No separate permit fees are issued in Pitt Meadows for the fire sprinkler system (Fig. 4.7).

Table 4.15 Community
sprinkler system costs—Pitt
Meadows, BC

| | Total cost | Sprinklered space | |
		Size (ft²)	$/ft²
House 1	$3,045	3,239	$0.94
House 2	$3,045	3,239	$0.94
House 3	$3,045	3,239	$0.94

Fig. 4.7 Pitt Meadows
sprinkler system cost
comparison, 2008–2013

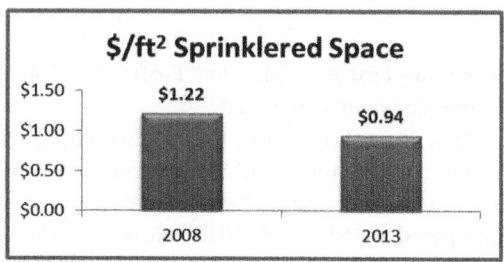

4.16 Prince George's County, Maryland

Prince George's County, MD is one of the selected communities included in both
the 2008 and 2013 studies. The county has a longstanding history with residential
sprinkler systems, dating back to 1987 when Prince George's County Council
approved the mandate of residential sprinkler systems. The law went into effect
January 1, 1992, requiring all residential structures, including single-family homes,
to install sprinkler systems per NFPA 13D standards. Maryland has since adopted
the 2009 IRC including the requirements for residential sprinkler systems, effective
January 1, 2011. Sprinkler systems in the county are typically installed by a
sprinkler contractor. Standalone systems are usually installed using CPVC piping.
The typical housing style in Prince George's County is two-story, 3,000 ft² homes
on basement foundations. Since 2008, the county has experienced a significant
amount of growth with 4,497 single-family building permits being issued according
to U.S. Census figures.[20]

For this study, three house plans were obtained from a local builder. Each of the
homes is custom built, two on basement foundations and one on a slab foundation.
The sprinkler system in the slab foundation home is supplied by a stored water
source, while the basement foundation homes are served by public water. Each
home has a standalone sprinkler system consisting of CPVC piping in the living
area and black steel in the unfinished areas. All three systems include a backflow
preventer. The area of sprinklered space for the three homes ranges from 1,700 to
2,400 ft². Total builder cost ranges from $2,640 to $4,153 ($1.10 to $2.44 per ft²
sprinklered area). A $75 permit fee is included in the total cost of each system
(Table 4.16).

---

[20] http://censtats.census.gov/

| | | Sprinklered space | |
Table 4.16 Community sprinkler system costs—Prince George's County, MD	Total cost	Size (ft^2)	$/ft^2
House 1	$2,640	2,400	$1.10
House 2	$2,995	2,000	$1.49
House 3	$4,153	1,700	$2.44

The smallest house (House 3) has the highest total and per unit area cost. Although there may be some economies of scale leading to higher unit costs for smaller homes, it is likely due in large part to the increased cost of systems supplied by a stored water source.

System cost increased between the 2008 study and the 2013 update. Average price per ft^2 of sprinklered space in 2008 was $1.00 while, in 2013, it is $1.68. Sprinkler system costs depend on a number of variables and when one changes, the cost figures can change significantly. In the 2008 study, all three homes analyzed were tract homes supplied by public water sources. The data for 2013 comes from three custom homes, one of which is served by a stored water source. In general, tract homes have lower design, installation, and inspection fees because sprinkler contractors tend to be familiar with the floor plans. Further, having one system out of three served by a stored water supply skewed the 2013 average.

4.17 Scottsdale, Arizona

Scottsdale, AZ has mandated residential sprinkler systems since 1986. Code requires that sprinkler systems be installed per NFPA 13D standards. It also requires garage and mechanical areas to be covered by the system. In the community, served by the Scottsdale Fire Department, standalone systems of CPVC piping are most common. Sprinkler systems are generally installed by a sprinkler contractor. There is a wide variety of housing styles in Scottsdale from tract homes around 2,000 ft^2 to custom homes over 6,000 ft^2. Slab foundations are predominant. Since 2008, 1,035 single-family building permits have been issued in Scottsdale according to U.S. Census figures.[21]

For this study, three house plans were obtained from a local sprinkler contractor in Scottsdale. All three homes are large custom homes on slab foundations ranging from 7,200 to 8,200 ft^2. Including garage and mechanical areas, sprinklered area ranges from 8,000 to 9,400 ft^2. All homes are connected to the public water supply.

---

[21] http://censtats.census.gov/

		Sprinklered space	
Table 4.17 Community sprinkler system costs— Scottsdale, AZ	Total cost	Size (ft^2)	$/ft^2
House 1	$9,200	9,400	$0.98
House 2	$8,700	8,200	$1.06
House 3	$8,500	8,000	$1.06

Each sprinkler system is a standalone system constructed of CPVC piping. Total cost to builder for the three systems ranges from $8,500 to $9,200. In terms of cost per sprinklered ft^2, the homes range from $0.98 to $1.06 (Table 4.17).

Appendix: All Community Data

Community and house plan	Area of sprinklered space (ft^2)	$/ft^2 Sprinklered space	Total cost
Addison, TX	7,045	$1.70	$11,972
Bakersfield, CA	4,013	$1.29	$5,180
Bakersfield, CA	3,216	$1.22	$3,930
Bakersfield, CA	3,537	$1.16	$4,115
Baltimore County, MD	3,150	$1.66	$5,225
Baltimore County, MD	3,100	$1.15	$3,550
Baltimore County, MD	4,800	$1.05	$5,050
Carroll County, MD	5,248	$1.33	$7,000
Carroll County, MD	5,729	$1.22	$7,000
Carroll County, MD	4,412	$1.13	$5,000
Cheatham and Davidson County, TN	4,800	$1.35	$6,480
Cheatham and Davidson County, TN	11,045	$1.15	$12,700
Cheatham and Davidson County, TN	2,400	$1.10	$2,640
Elk Grove, CA	2,004	$1.11	$2,229
Elk Grove, CA	3,358	$0.99	$3,332
Elk Grove, CA	4,258	$0.85	$3,600
Fort Collins, CO	3,600	$2.32	$8,365
Fort Collins, CO	4,100	$1.80	$7,385
Fort Collins, CO	5,500	$1.63	$8,975
Fort Collins, CO	6,600	$1.55	$10,200
Fresno, CA	2,399	$0.94	$2,252
Fresno, CA	2,647	$0.82	$2,181
Fresno, CA	2,090	$0.81	$1,695
Greenburgh, NY	8,500	$2.47	$21,000
Greenburgh, NY	6,000	$1.83	$11,000
Greenburgh, NY	4,500	$1.78	$8,000
Irvine, CA	6,240	$1.44	$9,008
Irvine, CA	2,570	$0.94	$2,408
Irvine, CA	3,112	$0.89	$2,784
Irvine, CA	2,830	$0.87	$2,466

(continued)

N. Partners LLC, *Home Fire Sprinkler Cost Assessment*, SpringerBriefs in Fire,
DOI 10.1007/978-1-4939-1083-0, © Fire Protection Research Foundation 2014

Community and house plan	Area of sprinklered space (ft^2)	$/ft^2 Sprinklered space	Total cost
Kenmore, WA	2,601	$1.82	$4,740
Kenmore, WA	2,498	$1.82	$4,540
Kenmore, WA	2,498	$1.82	$4,540
Lake County, IL	4,084	$2.08	$8,491
Lake County, IL	5,121	$1.83	$9,386
Lake County, IL	8,146	$1.71	$13,909
Montgomery County, MD	4,074	$1.26	$5,130
Montgomery County, MD	4,896	$0.92	$4,517
Montgomery County, MD	5,268	$0.92	$4,834
North Andover, MA	2,864	$1.47	$4,200
North Andover, MA	4,306	$1.44	$6,200
North Andover, MA	4,314	$1.11	$4,800
Pitt Meadows, BC	3,239	$0.94	$3,045
Pitt Meadows, BC	3,239	$0.94	$3,045
Pitt Meadows, BC	3,239	$0.94	$3,045
Prince Georges County, MD	1,700	$2.44	$4,153
Prince Georges County, MD	2,000	$1.50	$2,995
Prince Georges County, MD	2,400	$1.10	$2,640
Scottsdale, AZ	8,000	$1.06	$8,500
Scottsdale, AZ	8,200	$1.06	$8,700
Scottsdale, AZ	9,400	$0.98	$9,200

Printed by Printforce, the Netherlands